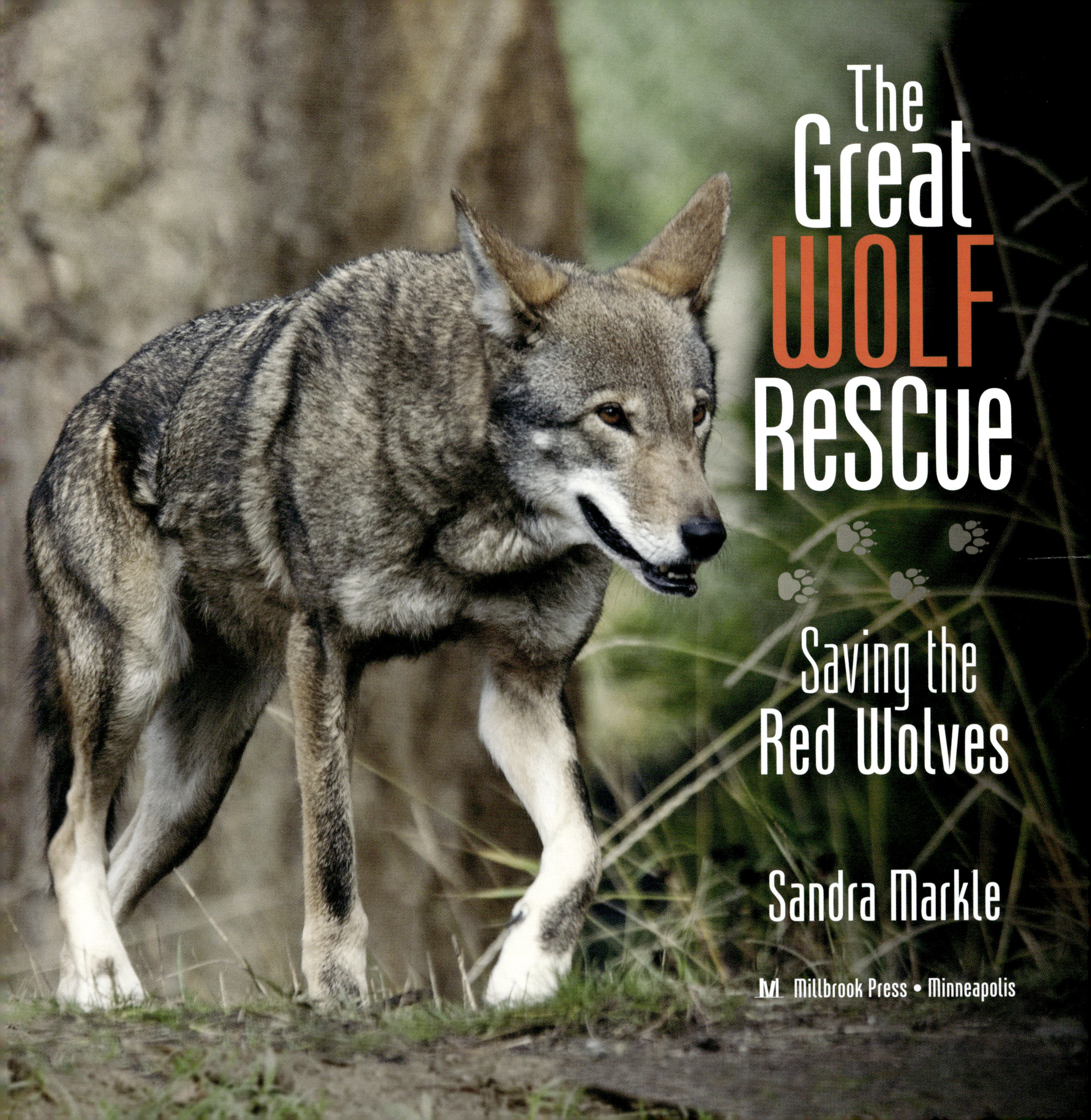

The Great WOLF ReSCue

Saving the Red Wolves

Sandra Markle

Millbrook Press • Minneapolis

For Rachel Knapton and all the children of Penn Yan Elementary School in Penn Yan, New York

Acknowledgments: The author would like to thank the following people for sharing their enthusiasm and expertise: Eric Albers, red wolf curator, Akron Zoo, Akron, Ohio; Rebecca Bose, curator, Wolf Conservation Center, South Salem, New York; Dr. Kristin E. Brzeski, Michigan Technological University, Houghton, Michigan; Dr. Joseph Hinton, Wolf Conservation Center, South Salem, New York; Maggie Howell, former executive director, Wolf Conservation Center, South Salem, New York; Chris Lasher, Red Wolf Species Survival Plan coordinator, North Carolina Zoo, Asheboro, North Carolina; LightHawk Conservation Flying coordinated by Esther Duke (western program director) and Audrey Ek-Psomas (eastern program director), Grand Junction, Colorado; Joe Madison, North Carolina program manager for Red Wolf Recovery Program, US Fish & Wildlife Service, Manteo, North Carolina; Regina Mossotti, director of animal care and conservation at Endangered Wolf Center, Saint Louis, Missouri; Frank Steeves, volunteer pilot for LightHawk, Appleton, Wisconsin; Dr. Ronald Sutherland, chief scientist, Wildlands Network, Salt Lake City, Utah; Dr. Bridgett vonHoldt, Princeton University, Princeton, New Jersey; Kim Wheeler, executive director of the Red Wolf Coalition, Columbia, North Carolina; Robert Wilcox, Museum of Life and Science, Durham, North Carolina.

A special thank-you to Skip Jeffery for his loving support during the creative process.

Millbrook Press™
An imprint of Lerner Publishing Group, Inc.
241 First Avenue North
Minneapolis, MN 55401 USA

For reading levels and more information, look up this title at www.lernerbooks.com.

Maps on pages 7, 16, 20 by Laura K. Westlund.

Designed by Viet Chu.
Main body text set in Metro Office. Typeface provided by Linotype AG.

Library of Congress Cataloging-in-Publication Data

Names: Markle, Sandra, author.
Title: The great wolf rescue : saving the red wolves / by Sandra Markle.
Description: Minneapolis : Millbrook Press, [2024] | Series: Sandra Markle's science discoveries | Includes bibliographical references and index. | Audience: Ages 9–12 | Audience: Grades 4–6 | Summary: "Red wolves had been hunted until only fourteen survived in the wild. So scientists hatched a daring plan—capture all of the wild red wolves and breed them in captivity. Discover the story of their recovery" —Provided by publisher.
Identifiers: LCCN 2023021171 (print) | LCCN 2023021172 (ebook) | ISBN 9781728467962 (library binding) | ISBN 9798765619308 (epub)
Subjects: LCSH: Red wolf—Conservation—United States—Juvenile literature. | BISAC: JUVENILE NONFICTION / Animals / Wolves, Coyotes & Wild Dogs
Classification: LCC QL737.C22 M3628 2024 (print) | LCC QL737.C22 (ebook) | DDC 599.773—dc23/eng/20230525

LC record available at https://lccn.loc.gov/2023021171
LC ebook record available at https://lccn.loc.gov/2023021172

Manufactured in the United States of America
1-51837-50464-10/3/2023

TABLE OF CONTENTS

A red wolf can hear sounds from 6 miles (9.7 km) away and detect scents from 1.7 miles (2.7 km) away.

STRANGER DANGER!

The female red wolf charged out of her den and ran through the forest. She didn't go far, though. As soon as she reached tall grass and tree shadows, she stopped. She had bolted when she heard and smelled a *human* coming close to her den. Fleeing was her way of leading this possible danger away from her ten-day-old pups. From her hiding spot, she listened and sniffed for clues that the intruder was staying away from her family.

At birth, red wolf pups are about the size of soft drink cans and have only a thin fur coat. When their mother leaves to find food, they huddle together to stay warm.

Why was a human near the red wolf's den? Would the pups stay safe?

Red wolves were named for the reddish color of their fur, which shows up strongest on their head and legs. Their scientific name is *Canis rufus*. *Canis* means "wolf" and *rufus* means "red."

START AT THE BEGINNING

Long ago, lots of red wolves lived in North America. They roamed from southern New York throughout the United States as far south as Florida and as far west as Texas. The red wolf was one of the main predators in the region, living alongside black bears and mountain lions in much of its territory. Native peoples, including the Cherokee, also shared the land with these wolves. They continue to consider them part of their cultural heritage, believing red wolves originally taught humans how to hunt.

But European people who moved into those areas saw the red wolves as a threat to themselves and to their livestock. During the 1800s and early 1900s, they shot and trapped red wolves in large numbers, decreasing their population. The settlers also destroyed the red wolf's habitat by clearing land to farm and build homes. Worse, they spread out and built towns that grew into cities. Eventually, red wolves struggled to survive where they had once thrived.

By the 1960s, red wolves were only seen roaming wild along the Gulf Coast of Texas and Louisiana. Even there, sightings were rare. In 1973, with worldwide concern about conserving wildlife growing, the United States established the Endangered Species Act. And the red wolf officially became an endangered species, a kind of wildlife whose population is so small it needs to be protected. The United States Fish and Wildlife Service (USFWS) became responsible for ensuring the red wolf's survival. To accomplish this, in 1973, they established the USFWS Red Wolf Recovery Program.

But leading up to that point, something else happened to shrink the red wolf population across its historic range. Coyotes moved into red wolf territory. And when a suitable red wolf partner wasn't available because the wolves sensed they were close relatives, red wolves mated with coyotes. Red wolves and coyotes are both canids, members of the dog family. They could mate and produce pups, but their offspring were hybrids, a mix of both species. And the more the number of hybrids increased the more the pure red wolf population diminished.

WHICH ONES ARE RED WOLVES?

With the wild red wolf population continuing to shrink, the Red Wolf Recovery Program team decided to launch a captive breeding program. They would catch all the pure red wolves and breed them in zoos and wildlife refuges. This plan would ensure that captive-bred pups would be pure red wolves, and the species would be kept from extinction (no more would exist). Once the captive population was large enough, some of those wolves could be reintroduced into protected areas within the red wolf's historical range to reestablish the species in the wild.

From 1973 through 1979, the Red Wolf Recovery Program team worked to trap all the remaining wild red wolves. Joseph Hinton, research scientist with the Wolf Conservation Center, said, "They trapped more than four hundred canids in Louisiana and Texas that were believed to possibly be red wolves." But which of these were pure red wolves suitable for launching the captive breeding program?

Red wolf-coyote hybrids can mate and produce pups, which are also hybrids. But the Red Wolf Recovery Program hoped to breed pure red wolves, such as the mother and pup shown here.

This one-day-old red wolf pup is one of the newest members of the captive red wolf population.

To preserve the red wolf population, hybrids needed to be excluded from the captive red wolf population. As hybrids pass along coyote and wolf genes, breeding more of them would not contribute to the goal of increasing the pure red wolf population. But in the 1970s, genetic testing wasn't readily available. So, the recovery team made a checklist for selecting animals for the captive breeding program. This list was based on what the team determined to be distinctive red wolf features, such as coat color and howl pattern.

From approximately four hundred captured and examined animals, just seven males and seven females were selected as pure red wolves to start the captive breeding program. Those fourteen animals were moved to the Point Defiance Zoo and Aquarium in Tacoma, Washington.

RED WOLF OR COYOTE?

The Red Wolf Recovery Program team checked for known red wolf traits and coyote traits to select animals believed to be pure red wolves.

COYOTE
COLOR
Usually grayish-brown fur on its body (though can vary from silver gray to black) with a black-tipped tail
SIZE
22 inches (56 cm) tall at its shoulder, 3 feet (0.9 m) long from nose to tail tip
EARS
Large, triangular ears
MUZZLE
Long and narrow
HOWL
Short rising and falling tone punctuated with yips, yaps, and barks
WEIGHT
Between 20 and 35 pounds (9 and 16 kg)

SAFE BUT CAPTIVE

In 1980 red wolves were officially declared extinct in the wild because no new sightings were being reported. That didn't mean there were absolutely none left living free. But the effort to recover more red wolves stopped. The Red Wolf Recovery Program's focus turned to the captive breeding program that was already underway.

As early as the late 1960s, some animals believed to be red wolves had been caught and put in zoos with the hope of eventually starting a breeding program. But by 1980, most of those animals had died. Upon closer inspection, the few remaining offspring of that zoo population were determined to be hybrids, so they were not included in the captive breeding program. And two of the fourteen wild red wolves (one male and one female) chosen for the breeding program were unable to reproduce. In 1984 the Red Wolf Species Survival Plan, a separate but key part of the Red Wolf Recovery Program, officially launched at the Point Defiance Zoo and Aquarium with six females and six males.

As of February 2023, there were 235 captive red wolves living in forty-nine facilities across the United States.

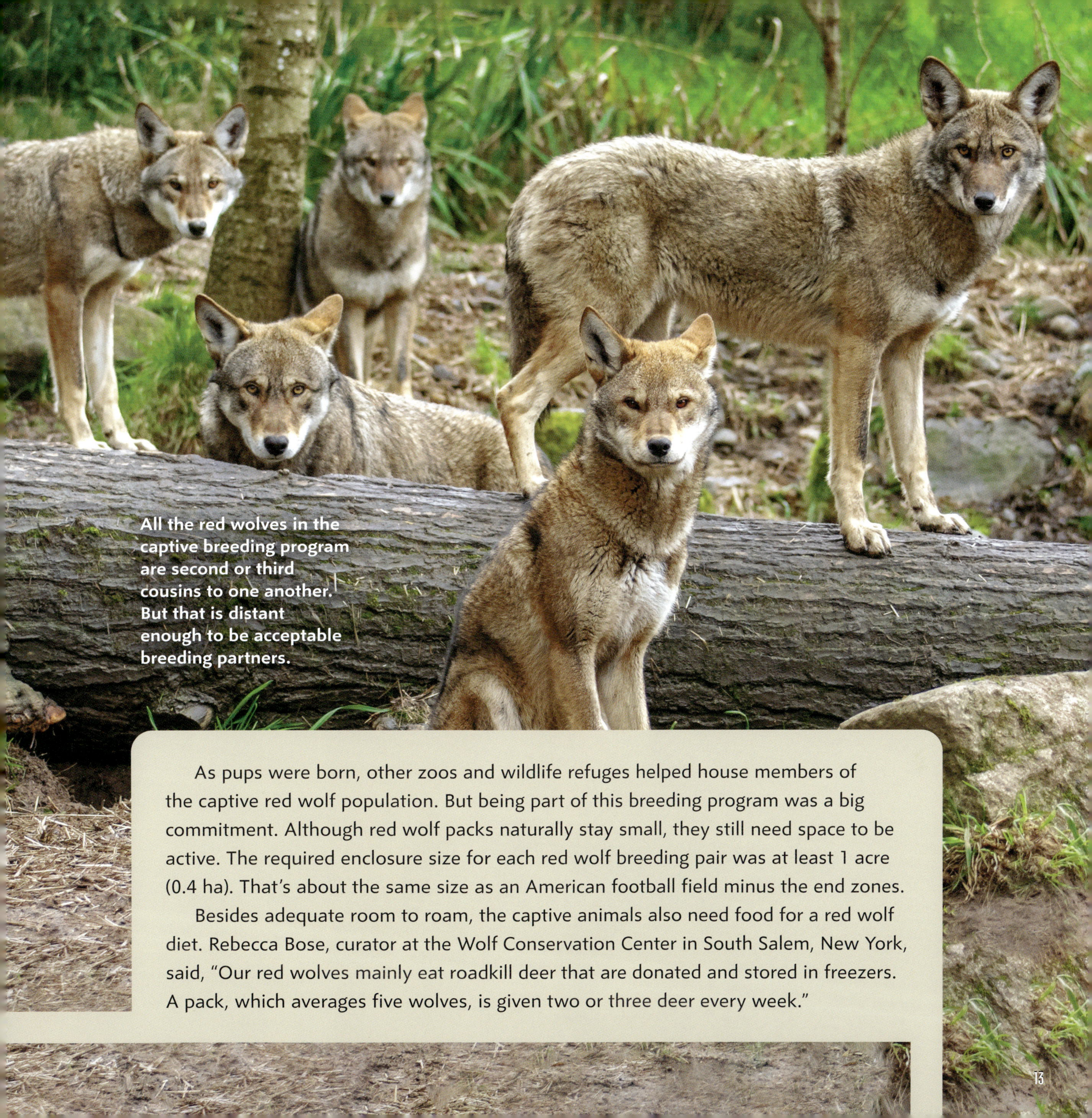

All the red wolves in the captive breeding program are second or third cousins to one another. But that is distant enough to be acceptable breeding partners.

As pups were born, other zoos and wildlife refuges helped house members of the captive red wolf population. But being part of this breeding program was a big commitment. Although red wolf packs naturally stay small, they still need space to be active. The required enclosure size for each red wolf breeding pair was at least 1 acre (0.4 ha). That's about the same size as an American football field minus the end zones.

Besides adequate room to roam, the captive animals also need food for a red wolf diet. Rebecca Bose, curator at the Wolf Conservation Center in South Salem, New York, said, "Our red wolves mainly eat roadkill deer that are donated and stored in freezers. A pack, which averages five wolves, is given two or three deer every week."

RED WOLF DATING PROGRAM

Like captive breeding programs for any kind of animal, the red wolf captive breeding program needed adult pairs to produce pups. And the captive red wolf population needed to remain as genetically diverse as possible. Genetic diversity is important so traits that can cause animals to be less healthy are not as likely to be passed on. Achieving both required careful planning and a little luck that the arranged mates would reproduce. Chris Lasher, coordinator for the Red Wolf Species Survival Plan, said, "Every year in July, the committee meets to decide how many new red wolves we have room to house. And we have an expert who tells us, if we want to expand our population by a certain number, how many pairs we need [for the breeding season] in the hopes of adding enough pups to reach that number."

To create red wolf breeding pairs that are as unrelated as possible, every red wolf has an identification number. That way the members of the Red Wolf Species Survival Plan team can easily identify each red wolf's parents, grandparents, great-grandparents, and even earlier ancestors.

Red wolves are ready to begin mating and birthing pups when they are about three years old.

Lasher explained, "We fly wolves all over the country to get males and females together before their annual breeding season [January through early March]. However, if we put thirty pairs together, we're only likely to have ten to twelve successful pair matches which produce pups." Still, that success rate was strong enough for the number of captive red wolves to steadily increase.

With genetic testing more readily available, scientists have discovered that red wolves have some coyote genes. So, an individual animal is considered a red wolf rather than a hybrid when its genetic makeup is at least 87.5 percent red wolf.

GOING WILD AGAIN

By 1984, the captive red wolves had already been reproducing and increased the population from twelve to sixty-three red wolves. Those working on the Red Wolf Species Survival Plan team decided it was time to consider returning red wolves to the wild. It would be the first time in North America that a large predator was set free to roam the land—eleven years before the 1995 reintroduction of gray wolves into Yellowstone National Park. But there was one *big* question: *Where should red wolves be reintroduced?*

The Red Wolf Recovery Program team chose Alligator River National Wildlife Refuge in eastern North Carolina, a part of the red wolf's historical range.

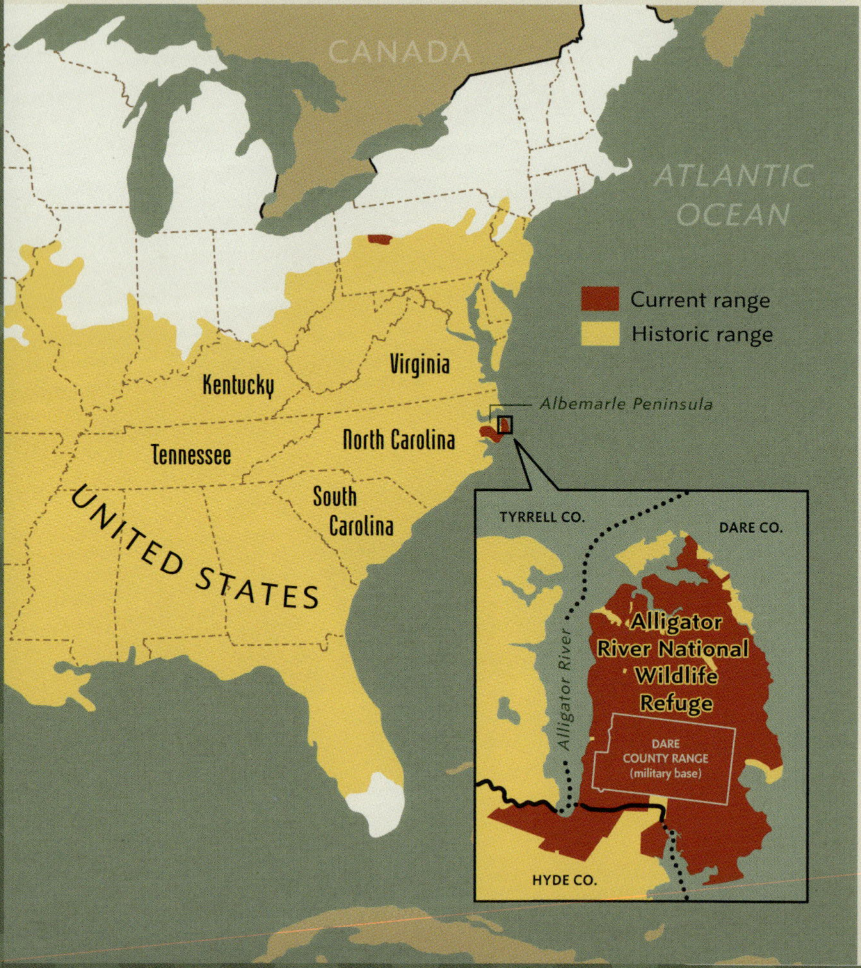

Another reason for choosing this area was its large size. Covering 120,000 acres (48,562 ha), Alligator River National Wildlife Refuge was big enough for many red wolf packs to claim home territories. Yet another plus was that the chosen release site was partly a peninsula. With the land surrounded on three sides by water, it would help restrict the released red wolves to their new home. And it would limit

Alligator River National Wildlife Refuge is on Albemarle Peninsula in North Carolina's Outer Banks.

human interference in the red wolves' lives. In addition, as far as the Red Wolf Recovery Program team's study of the area showed, no coyotes were living there to compete for prey or mate with the red wolves and produce hybrid pups.

In 1987 four red wolves—two males and two females—were released into Alligator River National Wildlife Refuge, starting the new wild red wolf population. Hinton explained, "First, they were put into acclimation pens in the planned release area. These were fenced areas about half the size of a house. And once a week USFWS biologists would toss in roadkill deer to give them a taste of their wild diet." After several weeks in the acclimation pens, the red wolves were released to roam free.

But the wolves were not completely left alone. Before they were released, the Red Wolf Recovery Program team fit the wolves with tracking collars to monitor their movements within the refuge. Team members also regularly hiked into the reserve to check on the red wolves. Everyone working on the Red Wolf Recovery Program and the Red Wolf Species Survival Plan was thrilled when, in 1988, both released pairs produced pups.

Each red wolf pair's territory covered from 12,800 to 51,200 acres (5,180 to 20,720 ha). Territory size depends on how much habitat provides enough food all year.

Red wolf pups are born in April or May. The mother may have one or as many as six pups in a litter—sometimes even more. At birth, each pup weighs about 1 pound (0.5 kg), and its eyes and ears are sealed shut.

At about two weeks old, a pup can see, hear, and walk. In addition to nursing (feeding on milk its mother produces), a pup begins to eat a little solid food. Adults bring up partly digested meat they have eaten to feed a pup. This food is easy for a pup to chew and helps introduce it to a red wolf's natural diet. At six months, a juvenile red wolf is nearly full grown and goes along on hunts.

At a year old, a young adult red wolf helps support its pack. It hunts and brings meat to the den site for its mother while she nurses the next litter of pups. Later, the young adult helps feed and watch over the growing pups. This is all on-the-job training for the future. At two years old, a young adult red wolf leaves its parents to roam on its own until it finds a mate, claims a territory, and starts its own family pack.

This red wolf is wearing a tracking collar to monitor where it roams.

The pups often chase one another and wrestle. Playing like this helps them grow stronger and develop skills they will need as adults.

So, based on the red wolf's life cycle, it took several years for the reintroduction program to naturally expand. That was when the original released pairs' pups matured and left to start their own families in newly claimed territories.

REINTRODUCED THEN RELOCATED

In 1989, with 105 red wolves in captivity, the Red Wolf Recovery Program team decided to launch a second reintroduction. The long-term goal for the program was to maintain 330 red wolves in the captive population while building up the wild population to at least 120 red wolves. But suitable sites for wild red wolf populations were limited within the eastern United States—the red wolf's historical range. People had changed much of that area into cities connected by a network of highways.

The site the Red Wolf Recovery Program team selected for a second red wolf reintroduction was the Great Smoky Mountains National Park. Covering 522,427 acres (211,419 ha) spread across North Carolina and Tennessee, this was one of the largest remaining wilderness areas within the red wolf's historical range. In 1992 a pair of red wolf adults were released there along with their five pups. And that year, two red wolf adults and their four pups were released into another area in North Carolina—Pocosin Lakes National Wildlife Refuge.

USFWS biologist Chris Lucash releases a captive-born red wolf to roam wild.

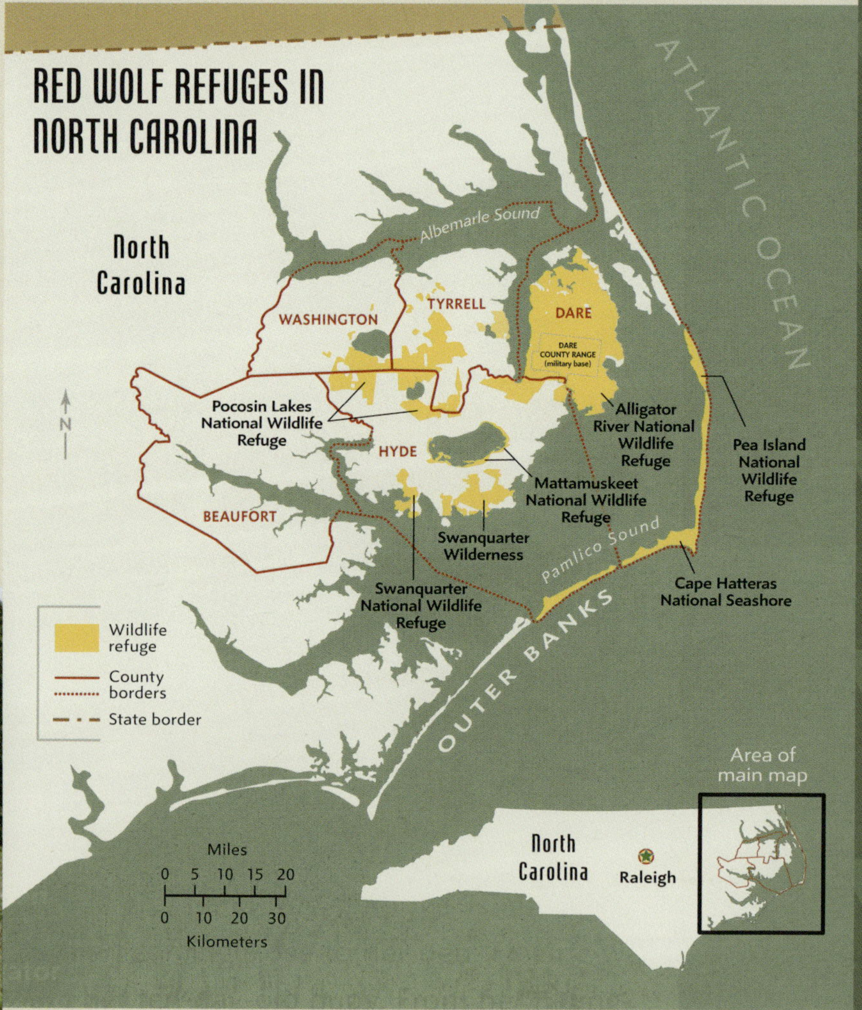

During the 1990s, the North Carolina area set aside for the red wolves expanded to include both federal and private land in five counties. All the red wolves roaming that area are considered part of the Eastern North Carolina Red Wolf Population.

Over the next seven years, thirty-seven more adult red wolves were released in the Great Smoky Mountains, and thirty-three pups were born there. But as that red wolf population grew larger, it became increasingly at odds with its human neighbors. Since most of the area wasn't fenced, the red wolves didn't stay inside the park. Several red wolves roamed close to farms and homes where they were shot or trapped. So, in 1998, the Red Wolf Recovery Program team captured and removed all red wolves from the Great Smoky Mountains National Park. Some were sent to zoos or wildlife refuges to be part of the captive population. Others were released to join the wild populations at the North Carolina sites.

Despite the failure of the Great Smoky Mountains National Park population, the rest of the red wolf reintroduction program seemed to be doing well. But was it really? Only tracking and observing red wolves living wild would tell.

Due to conflict with their human neighbors, some of the red wolves roaming the Great Smoky Mountains National Park were moved to the Eastern North Carolina Red Wolf Population, which includes Alligator River and Pocosin Lakes National Wildlife Refuges.

WATCHING OVER WOLVES

Tracking collars let the Red Wolf Recovery Program team check up on the reintroduced red wolves. These collars provide very high frequency (VHF) monitoring to pinpoint a red wolf's location and observe it. Joe Madison explained, "Monitoring VHF is typically done by driving along roads . . . with an antenna on top of the vehicle. That detects the [collared] animal in any direction a decent amount of distance away from the vehicle. Then once you detect an animal you get out of the vehicle and use a handheld antenna. That lets you determine the direction of the signal from the red wolf's collar and provides a general idea of how far away the animal is. When you keep doing this in multiple spots, you can finally locate the red wolf."

Some tracking collars include Global Positioning System (GPS) monitors. GPS tracking shows a red wolf's general movements as it roams and reveals where it stays for any period of time.

By 2011 as many as eighty-nine of the red wolves wore tracking collars to monitor their activities.

Red wolves often travel along roads because it's easier than walking through dense underbrush in the woods. But that puts them at risk of being struck by vehicles.

With the help of camera traps, the Red Wolf Recovery Program team observed the red wolves roaming wild. The team set up the camera traps in areas where they found plenty of tracks and scat (waste droppings)—clues the red wolves frequented those places. The camera traps captured photos or videos of wild red wolf behavior when no humans were around. Madison said, "We typically put cameras up in known red wolf use areas to monitor their health, see who is with whom [forming mated pairs], and check on pup survival."

All the monitoring efforts paid off, revealing previously unknown behavior for this rare wolf species. The Red Wolf Recovery Program team discovered young adult red wolves without mates roamed larger areas than those that had formed a family pack. And the team learned all wild red wolves ranged over larger areas during the winter than during the summer.

Seeing new pups gave the Red Wolf Recovery Program team hope that the reintroduced red wolf population was strong and growing.

FROM GOOD TO TERRIBLE

By 2012 the Red Wolf Recovery Program team recorded the target population size of 120 red wolves in the North Carolina reintroduction sites. So, from 2012 through 2015, the team suspended releases of captive adult red wolves. And in 2015, the USFWS officially ceased adding captive red wolves to the wild population. But the wild red wolves still faced serious challenges.

Just as when European settlers first came to the area, many people living where the newly established red wolf population roamed considered them a threat. And shooting any red wolf caught killing livestock remained legal. Another challenge for the reintroduced red wolf population was coyotes. They had arrived in the 1990s, claiming territories as the native red wolf population diminished. And they never left.

The reintroduced red wolf population faced the same conflict with coyotes as their ancestors. Coyotes hunted the same prey animals and were competition for mates. As before, a young red wolf unable to find another red wolf suitable to be its partner might accept a coyote. The coyote-wolf pairs produced hybrid pups that some people call coywolves. And each new litter of hybrid pups reduced the chance for a pure red wolf to find and choose another red wolf for its mate.

While coyotes (*shown*) are natural competitors for red wolf habitat and prey animals, they have much smaller home ranges than red wolves. They are also more adaptable to living near areas humans have developed.

The North Carolina Zoo was excited about pup fostering. Being raised by a wild mother and her pack was the best way for captive-born pups, such as these, to grow up ready to survive living free.

DESPERATE TIMES REQUIRE DARING ACTION

When conservation groups learned of the shrinking reintroduced red wolf population, they filed a number of lawsuits to require the USFWS to release more captive red wolves. But year after year the legal struggle failed, and the wild red wolf population continued to decline due to habitat competition, being shot, or being hit by cars.

By January 2021, the wild red wolf population was in serious trouble. A court order restarted the release of captive red wolves into the Eastern North Carolina Red Wolf Population. But only eleven red wolves remained in the wild population. These were all collared and tracked. Alhough six of the eleven were females, none had bred and produced pure red wolf pups in the three years they were monitored: 2019, 2020, and 2021. Instead, they had mated with coyotes and either failed to produce pups or produced hybrid litters.

In 2021, to jump-start rebuilding the wild red wolf population, the Red Wolf Recovery Program team released four captive adult red wolves (two males and two females). The team also decided to attempt something daring if they had the opportunity. They would do pup fostering. This meant placing captive-born red wolf pups into a wild red wolf mother's den for her to raise along with her pups—even if she had a hybrid litter.

Madison explained, "We'd been going into the red wolf dens since the late 1990s to count the pups and put microchips in them, the way people chip pets, for identification purposes. When [we] first did the chipping, we were worried human handling could cause the pups to be abandoned by their mother, and we'd have to rescue them for captive rearing. But the mother went right back to her pups."

The first pup-fostering attempt had happened in 2002, and it worked! The Red Wolf Recovery Program team then continued successfully pup fostering during breeding seasons until 2015. But pup fostering ceased when the team halted red wolf reintroductions.

In 2021, with the number of pure wild red wolves in the reintroduced population so small, the situation was desperate. Could pup fostering be a way to speed up increasing the pure red wolf population in the wild? After all, pups growing up learning how to survive wild and free had the best chance of becoming successful adults.

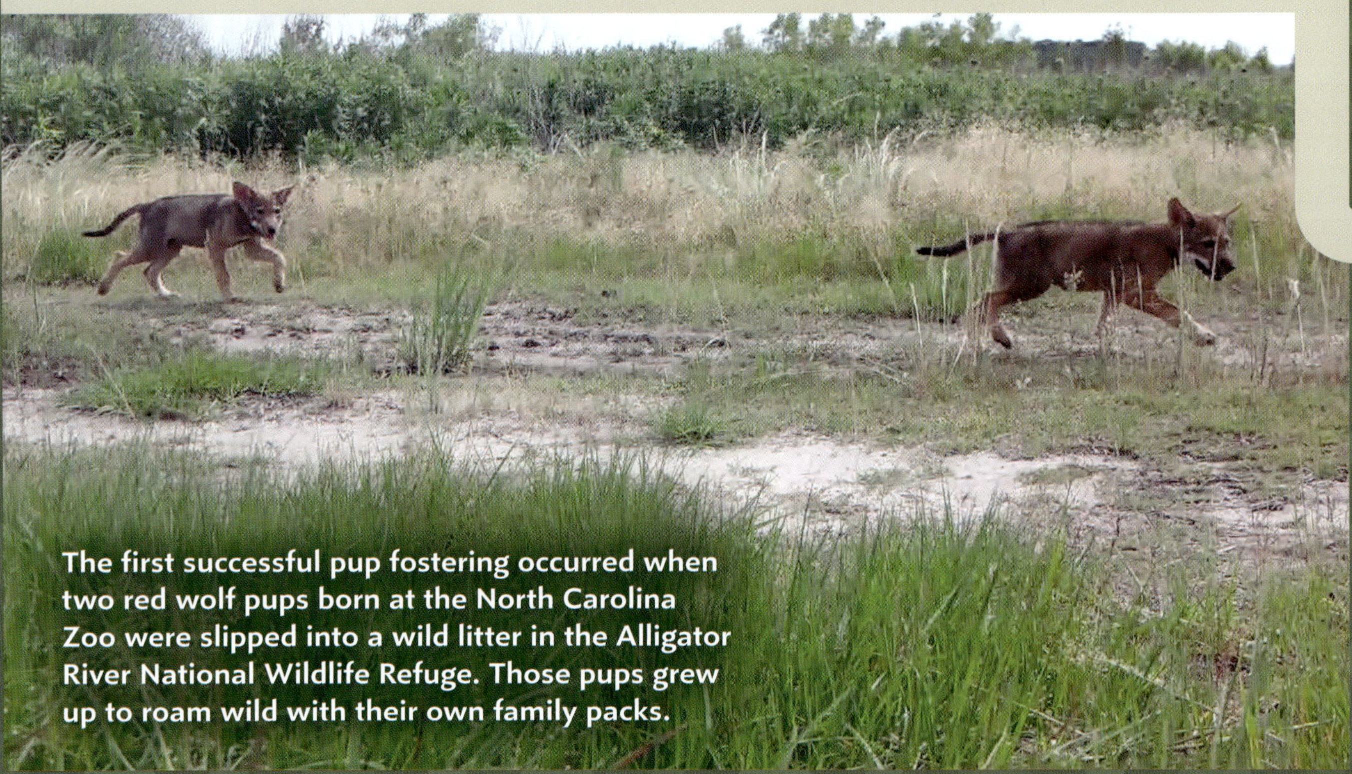

The first successful pup fostering occurred when two red wolf pups born at the North Carolina Zoo were slipped into a wild litter in the Alligator River National Wildlife Refuge. Those pups grew up to roam wild with their own family packs.

Unless they lose their partner, wild red wolves mate for life. But to maintain genetic diversity in the population, captive red wolves are moved around to create different mating pairs.

Madison said, “We got the idea to try pup fostering again [in 2021] when [female] red wolf 2197’s VHF collar showed she was staying at a den site. She was five years old and had already successfully raised pups. So, though now her mate was a coyote, she could surely foster captive-bred red wolf pups.” Adding to his hope for this pup-fostering opportunity, Madison knew a red wolf pair at the North Carolina Zoo had recently produced pups.

The key was going to be the age of 2197’s pups. From past pup-fostering efforts, those working with the red wolves had learned two important lessons. First, for introduced pups to be accepted by a wild mother, the pups needed to be very close in age to her own litter. Second, the introduction of the captive-born pups had to happen while they were between ten days to two weeks old. The team had learned that was when a red wolf mother was most likely to accept introduced pups. And ten days old was the earliest captive-born red wolf pups could skip nursing long enough to be transported to their new mother.

When Joe Madison crawled into 2197's den to check for pups, he was excited to discover she did have a litter. But there was a *big* problem. The pups born at the North Carolina Zoo were too much older than 2197's litter to be successfully pup fostered.

Madison quickly reached out to Lasher, the coordinator for the captive population. Lasher contacted all the facilities with red wolf breeding programs to see if any other captive red wolves had recently given birth. They were in luck! On April 22, 2021, a red wolf mother nicknamed Juno gave birth to a litter of eight pups at the Akron Zoo in Ohio. Those pups were a near perfect match in age to 2197's pups. And Juno's litter was unusually large so rehoming some of her pups would help her. Plans were made to pup foster four of Juno's pups.

Look how close in age 2197's pups appear to be to those born at the Akron Zoo.

TIME TO GROW UP WILD

Between the first pup fostering in 2002 and 2015, twenty-two captive-born red wolf pups placed into dens were successfully raised by wild mothers. But this was the first pup-fostering attempt in a long time. So, on May 1, 2021, when it was time for the pup-fostering attempt, everyone was both excited and a bit anxious.

For the trip, the red wolf pups were put on a towel in a soft carrier—the same kind people use for dogs (*top right*). The team placed hand warmers under one end of the towel so the pups had the option to cuddle there to warm up or spread out to cool off.

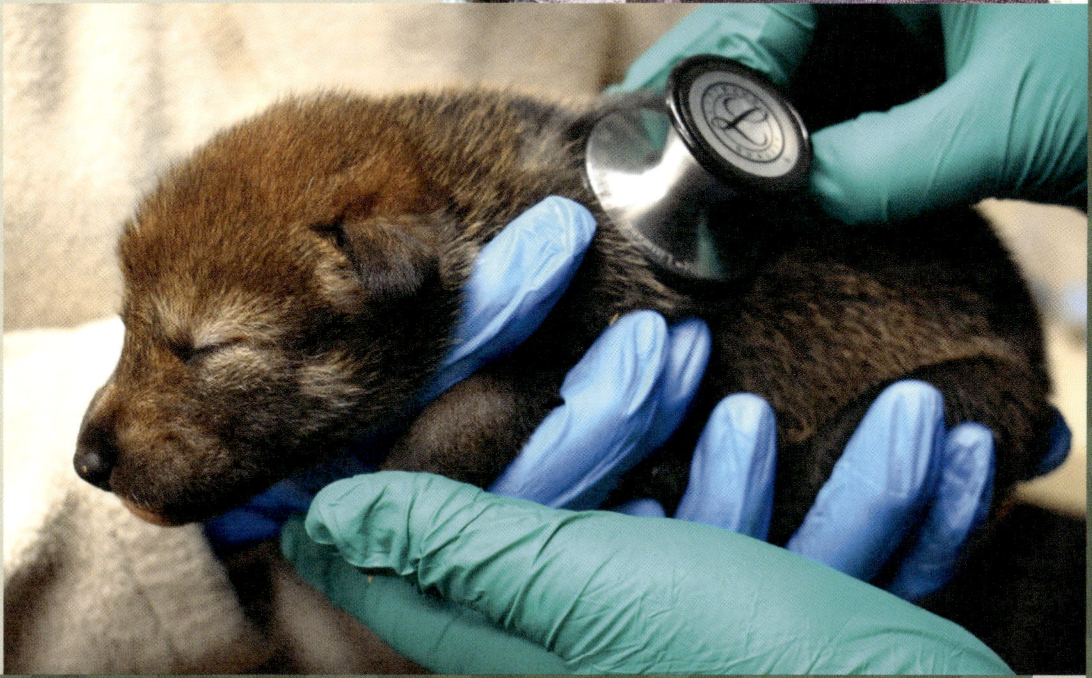

Before they left the Akron Zoo, a veterinarian checked the pups (*bottom right*). The two largest females and the two largest males (pups 2354, 2356, 2359, and 2361) were chosen for fostering because they had a better chance of surviving due to their size.

Eric Albers (*above*) took the red wolf pups in their travel carrier on board the plane and watched over them during the flight.

Frank Steeves flew his small Kodiak plane from his home in Appleton, Wisconsin, to Akron, Ohio, to pick up and transport the red wolf pups to North Carolina. Steeves regularly volunteered his time and plane to fly for LightHawk, a nonprofit organization that provides flights to help with disaster rescue missions and benefit endangered animal populations. He picked up the pups from the zoo, along with Eric Albers, who went along to watch over the pups. Then it was a less than two-hour flight to reach Plymouth Municipal Airport in North Carolina. Madison was waiting there to carry the pups to 2197's den.

Joe Madison (*left*) rubbed each of the captive-born red wolf pups against members of 2197's litter to make them smell familiar and help the foster mother accept them as part of her family.

Although a red wolf's jaws are strong enough to crush bones, the mother carries her pup so gently her teeth don't prick its skin.

THIS IS NOT THE END

When she sensed the human scent fading, the female red wolf cautiously returned to her den. With no intruder in sight, she rushed to the entrance and crawled inside. It was dark, but she didn't need to see to find her pups. And the instant they smelled her, they began a chorus of squeaky whimpers. The female pushed her nose into the warm, furry bunch of pups, sniffing and licking each one. Something was different, and she could smell the lingering hint of *human* here. Still, she gave in to her babies' pleas, settled onto her side, and let the pups nurse.

The Red Wolf Recovery Program team had suspected 2197 would move her pups after a human was inside her den, and she did. Red wolves often do that when they feel their den site is no longer safe. But the team had a good idea of where she went. So they set up camera traps in that area.

After weeks of anxious waiting, one of the camera traps finally captured a short video of 2197 with her family. This was only a glimpse, but it showed the red wolf mother had accepted the foster pups and was caring for them. Madison said, "Now, we need to get more pure red wolf breeding pairs out there and do some more pup fostering."

In 2022 ten more captive-bred red wolf adults were released into the Eastern North Carolina Red Wolf Population. Unfortunately, that attempt to increase the wild population failed. Three females and three males died either from gunshot wounds or being struck by vehicles. One female went missing, and her fate remains unknown. The final three males were recaptured and returned to captivity after they had to be repeatedly relocated from roaming close to people's homes. But there is good news.

The pup-fostered youngsters were growing up, adding four more red wolves to the wild population.

In 2022, for the first time since 2018, a wild red wolf pair (female 2225 and male 2323) produced a litter. Those six pups (four females and two males) were closely monitored as they grew up. Though red wolves are still primarily living in captivity, there is hope for North America's native wolves. That is because red wolf pups—some born wild and some fostered from captivity—are growing up to roam free.

And people are continuing the hard work needed to help red wolves have a wild future.

Although they are growing up in captivity, these four red wolf pups are adding to the hope that red wolves will survive.

A Note from Sandra Markle

There was a time in 2020 during the COVID-19 pandemic when I was at home all the time. I even had my groceries delivered. I thought back on that time of living a safe but limited life when I learned captive red wolf family packs, on average, are given a space of about an acre (0.4 ha) at wildlife centers and zoos. Living in such a restricted area has been true for all captive red wolves since their removal from the wild. Just like me in my quarantined space, they were safe, but their lives were limited. Wild red wolves usually roamed areas that at their smallest were about the size of Manhattan Island in New York. There they searched for and caught their own food. That was undoubtedly difficult at times, but it also challenged them and let them select what to eat. They could also choose their own mate. In captivity, food could be counted on, but it was delivered and so was a mate whose stay was limited. Imagine the difference!

I understand from my quarantined experience that while living in captivity provides safety, it is not the same as freely experiencing the world. I will always remember the marvelous feeling I had when I finally escaped my safe isolation at home, went out into the world, and gathered with friends. Surely, we can spare some of the world that is still wild and where red wolves once lived so they too can roam and experience all it means to live free.

Did You Know?

A red wolf pup can howl when it is about three weeks old.

A red wolf always walks and runs on its tiptoes.

A red wolf can run as fast as 30 miles (48 km) per hour and can travel 20 miles (32 km) in a day.

An adult male red wolf is bigger than a female. It weighs as much as 80 pounds (36 kg).

Although a red wolf usually eats prey animals, including rabbits, rodents, and deer, it also eats insects and berries.

Glossary

adult: a red wolf that is at least a year old. An adult is able to reproduce at about three years old.

biologist: a scientist who studies living things

climate change: major long-term changes in expected weather patterns

den: a hidden home, such as a burrow, where the female red wolf gives birth to her pups

endangered: a species that is at risk of becoming extinct

extinct: gone forever, no longer having any alive

habitat: the natural home environment of a plant or animal

juvenile: a young red wolf from about three months old until about a year old. This is a period of rapid growth and the beginning of joining adults for on-the-hunt training.

litter: a group of red wolf pups born at one time to a mother

nurse: to feed on milk from its mother's body

pack: a family group of red wolves that live and hunt together, usually made up of the parents, their current pups plus young adults until they are about two years old

predator: an animal that hunts, captures, and eats other animals

prey: an animal that a predator catches to eat

pup: a young red wolf from birth until about three months old

pup fostering: introducing captive-born wolf pups into a wild litter to be raised by a foster mother

territory: an area where a red wolf pack lives that has food, water, and a safe den site for the female to give birth to pups

Source Notes

8 Joseph Hinton, telephone interview with the author, July 12, 2021.

13 Rebecca Bose, telephone interview with the author, July 27, 2021.

14 Chris Lasher, telephone interview with the author, July 15, 2021.

15 Lasher.

17 Hinton, interview.

22 Joseph Madison, telephone interview with the author, November 11, 2022.

23 Madison.

28 Madison, telephone interview with the author, July 2, 2021.

29 Madison.

34 Madison.

Find Out More

Defenders of Wildlife: Captive-Born Wolves—With the Spirit of the Wild
https://defenders.org/blog/2021/08/captive-born-wolves-spirit-of-wild
Don't miss the video linked from this site that beautifully shares the history and future of red wolves.

Markle, Sandra. *On the Hunt with Wolves*. Minneapolis: Lerner Publications, 2023.
Investigate gray wolves, and compare them to red wolves.

Red Wolf Review Gallery
https://www.redwolfreview.info/galleries.html
Read an amazing collection of information about red wolves.

Red Wolves at North Carolina Zoo
https://www.youtube.com/watch?v=-2U4mY_pTDU
This is a virtual visit to the red wolves at North Carolina Zoo and to other red wolves.

Untamed Science: Red Wolves
https://www.youtube.com/watch?v=iyqm0iplWJI
Watch a great video introduction to red wolves and their reintroduction into the wild. Note: The number of red wolves listed for the current captive and wild populations shared in each of the videos is the number during a specific year. That total population number changes annually.

US Fish and Wildlife Service Red Wolf Recovery Program
https://fws.gov/project/red-wolf-recovery-program
This official site is for more advanced readers. But don't miss the year-by-year summary of red wolf releases and information about the wild population.

Index

Photo Acknowledgments

Note: Some images simulate actual events when it was impossible to photograph events unfolding in real time.

Image credits: FLPA/Alamy, p. 1; Prisma by Dukas Presseagentur GmbH/Alamy, p. 4; US Fish and Wildlife Service, pp. 5, 17, 18, 20 (left), 22, 23, 28, 30 (top), 32 (bottom); Mark Newman/Getty Images, pp. 6, 10, 14; Josh Lewis, pp. 9, 24; Ranger Robert Wilcox, Museum of Life and Science, pp. 8, 15, 33, 34, 35; bgsmith/Getty Images, p. 11; Austin Tucker/500px Prime/Getty Images, p. 12; karen crewe/Getty Images, p. 13; Christian Heinrich/Getty Images, p. 19; Matthew Cuda/Alamy, p. 21; Raimund Linke/Alamy, p. 25; Courtesy of the North Carolina Zoo, a division of the North Carolina Department of Natural and Cultural Resources, p. 26; Daniel Orchard/Getty Images, p. 29; Akron Zoo, pp. 30 (bottom), 31 (bottom), 32 (top); Regina Mossotti - Wildlife Biologist, p. 31 (top); Skip Jeffery Photography, p. 36; FLPA/Alamy, p. 37; design elements: Esra Sen Kula/Getty Images.

Front cover: Mark Newman/Getty Images; Esra Sen Kula/Getty Images.

Back cover: Marcia Straub/Getty Images.